The Universe Soup Outer Space Magic Journal

The Universe Soup OUTER SPACE MAGIC Journal #1
Written & Designed by Laura S. Woodmansee
ISBN-13: 978-0-9863172-9-3
Published in 2018 by Universe Soup.

OUTER SPACE MAGIC JOURNAL
Keep in touch on social media @LauraWoodmansee
Join our email list for goodies! UniverseSoup.com

UniverseSoup

This journal belongs to:

Outer Space Magic

*The world is
made of words,
and if you know the
words the world is
made of, you can make
of it whatever you wish*

-TERENCE MCKENNA

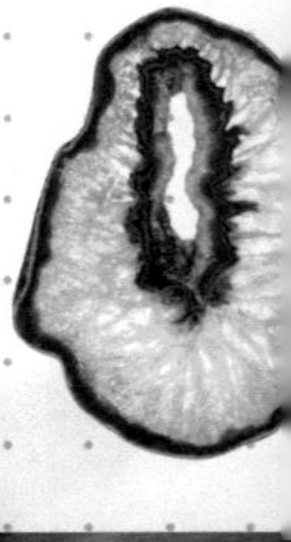

Outer Space Magic

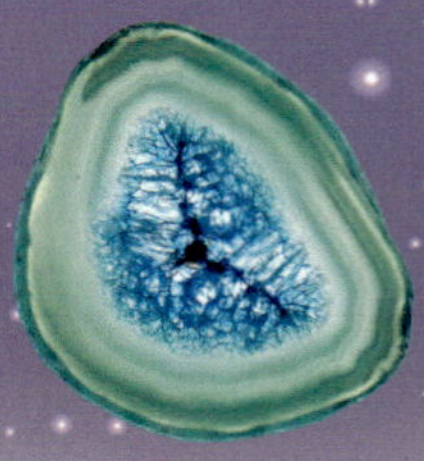

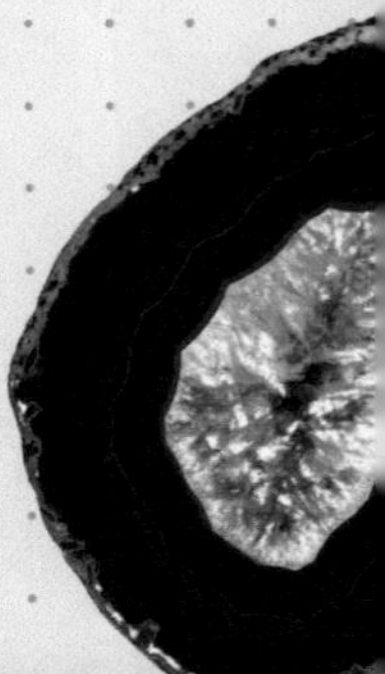

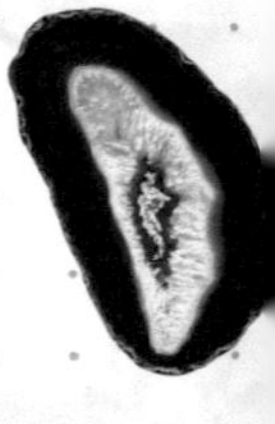

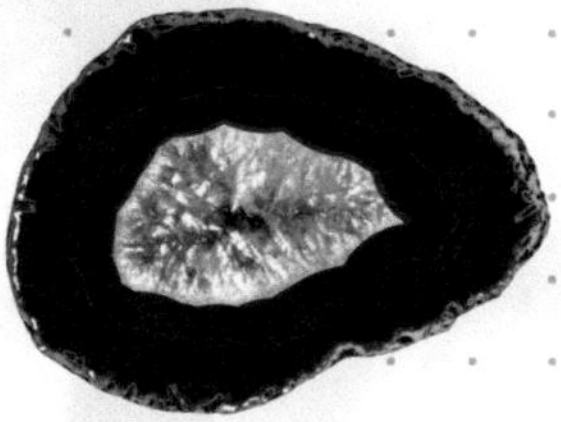

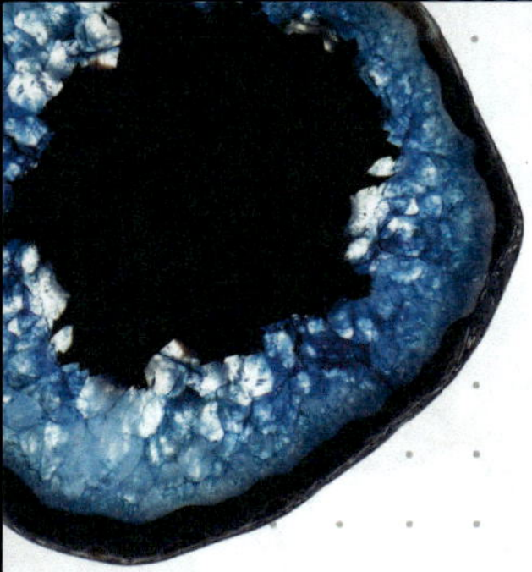

Cosmic
Outer Space Magic

Made in the USA
Monee, IL
05 November 2026

67381715R00026